BEGINNING FIELD BIOLOGY

BEGINNING FIELD BIOLOGY

by

O. N. BISHOP
B.Sc. (Bristol) B.Sc. (Oxon.)

HARRAP LONDON

First published in Great Britain
in 1963 as part of Beginning Biology
by GEORGE G. HARRAP & CO. LTD
182–184 High Holborn, London WC1V 7AX

© *O.N. Bishop* 1963, 1975

1SBN 0 245 52570 X

Printed in Great Britain by Biddles Limited, Guildford

CONTENTS

INTRODUCTION

Most of your studies of biology will have taken place indoors, in the laboratory, because this is the most convenient place for performing experiments and making observations. But this is not enough. At school you meet many people each day and you may get to know quite a lot about them — their names, what they look like, how clever they are at lessons and games, and so on. As long as you meet these people only at school you will never get to know them really well. To know people really well you must go and see them at home. The same is true of plants and animals — meeting them in school will tell you a lot about them but not all. To know more, you must visit them at home also. 'Field studies' is a term which means going to the homes of animals and plants to discover how they live in their natural surroundings. When used in this way the word 'field' covers not only a field but also a wood, a pond, a heath, a hedgerow, or the seashore, in fact, anywhere out-of-doors.

If you wish to do some field studies, it is best to spend a few lessons on them each term throughout the year. Then you will get to know about living things at all seasons of the year. The best time to start is early spring, when the main activity of the year is just beginning. You will be able to follow what happens during the spring and summer, notice what happens in autumn, and finally discover how plants and animals survive the more difficult months of winter, ready to begin again next spring.

This is a work-book and you will find plenty of ideas on what to do. The exact details of what you do depend a lot on the part of the country you live in and how easy it is for you to get to a suitable place. Those who live in rural areas will experience no difficulty; there will be woods, ponds, streams, and hedgerows in plenty. Those who are at school in the middle of a big city will find it more difficult to carry out field studies but there is plenty that can be done in a garden or in a park. There are many sorts of field work that can be done, partly depending on local circumstances and partly

7

depending on your own preferences and interests. In the first section you will find *suggestions* for field work. You may choose any one or more of these or perhaps, having got some ideas from the suggestions, you will be able to devise problems and projects of your own. When you have decided what to do, refer to the Methods Section to find out how to set about it. These methods will help you to get the most out of your work and find the answers to some of the problems. The answers to the problems are *not* in this book; they are there in the field — go out and find them.

THINGS TO DO — PROBLEMS AND PROJECTS

(1) A COMPLETE STUDY OF ONE SPECIAL HABITAT

A habitat is a place with a particular kind of vegetation and inhabited by certain types of animals. Examples of habitats are a wood, a pond, a stream, a heath, a moor, a river bank, a spring, the seashore, a hedgerow and a field. If you live in the country you will be able to choose from several of these. Some other habitats which are found both in country and in town are just as suitable for field studies: for example, a garden (particularly a weedy, overgrown one), a public park, a patch of waste ground, a building site, the school playing field, a garden rockery, a water butt, an old wall, a garden rubbish-heap.

It is also possible to make your own habitat for study. You will find some instructions under section 4 of Methods, p.30.

This sort of work is best done by a whole class, split into groups. Each group is responsible for one particular section of the work. Here are some things to do in your chosen habitat:

(*a*) Keep a diary in which you record events in the habitat each month. Note the flowers as they come out, the animals which live there and what they do at different seasons, the changes which occur in autumn, and the ways in which the living things survive the winter.

(*b*) Make a map, showing the main features of the habitat and any other special features (such as the position of rabbit warrens, nests, etc.) which are of interest to you (see Methods 1, p. 25).

(*c*) Make a list of all the plants found in the habitat. Record the name (see Methods 11, p.56), the frequency (see Methods 6 (*c*), p.45), the date of flowering, the date when fruits are first ripe and the methods by which the seeds or fruits are carried away from the parent plant in order to grow elsewhere.

(*d*) Make a list of all the animals found in the habitat (see Methods 11, p.56). List them under various headings, according to where they are found. For example, in a wood you might have the following headings: in the soil, in the leaf litter on the soil surface, on the plants (which plants?), feeding inside the plants (leaf miners, etc.), in rotten tree trunks and bark, underneath stones. In a pond the headings might be: *on* the water surface, *in* the water just below the surface, in deeper water, on the bottom, in the mud, under stones, on rocks and stones, on the water plants. You will be able to make up your own list of headings to suit your particular habitat. Do not forget to include in your list the larger animals – birds and mammals (see Methods 2 (*a*), p. 27, for use of the Longworth trap).

(*e*) When you have finished (*d*) look at your lists carefully. Why are some animals found only in one place? Is their food there? Is it a place where they can reproduce easily? Is there protection from the weather and enemies? Any other reason? For each of the animals on your list, try to find a reason for its being where it is. This can lead you to interesting work in devising some experiments to find out if your ideas are correct. For example, water snails are often found on stones which are covered by a thin film of algae; why are the snails there? Do they simply find it easier to move on stones or to stick to them, or do they eat the algae for food? Devise an experiment to answer this question. Do the experiment. What are your conclusions? There is another place in the pond where snails are often found; why are they there as well as on the stones?

This sort of information, especially if you add information from your own experiments, will make your list much more interesting. Illustrate your list by drawings of several of the animals.

(*f*) Study several of the animals in your list and find out what they feed on (see p. 13). Try to link them together in food chains. Draw your food chains on a large sheet of paper and illustrate with drawings of the animals concerned.

(*g*) If you have studied two habitats – say, a wood and a heath, or perhaps two different woods or two different ponds – compare the lists of plants and animals in each habitat. Are there any differences? Why? Remember that if you studied the two habitats at different times of year this may account for some of the differences in the lists. Differences may also be due to differences in soil composition (see Methods 10, p. 49), drainage, the presence or absence of another animal or plant on which the animal feeds, the presence or absence of suitable shelter, differences of light, wind, or humidity (see Methods 7 and 8, pp. 47-9). Carry out observations and experiments to check whether or not any of these differences can account for the differences in your two lists.

(*h*) In *one* habitat you may notice two or more distinct areas. If you do, study each area separately and make lists of plants and animals for each. Try to account for the differences. For example, areas in a wood which are pre-dominantly shaded will differ from those which receive more light (see Methods 6(*b*), p.37 and Fig. 8). On a bank or wall, there may be differences which will affect the plants and animals to be found at different heights. In a pond or stream, shade, water temperature, and currents can have effects. On the banks of a stream or beside a spring, the water content of the soil can vary a lot at different places and this can affect plants and animals. Underneath trees or in marshes the humus content of the soil is important (see Test 5, p.54).

(*i*) Study a hedgerow, preferably one that runs east-west. List the plants and animals as in (*c*) and (*d*). Make a special study of the climbing plants; write an illustrated account of them, classifying them under their various methods of climbing. Make separate lists of those animals and plants found on the north side of the hedge and those found on the south side. Can you explain the differences in these lists? If there is a ditch beside the hedge, include this in your study. The water in it may help to explain some of the differences between the two sides of the hedge. Do not forget to include birds nesting in the hedge and small mammals which may be living in the hedgebank. It is a good idea to illustrate your account by a profile diagram (see Methods 6(*b*), p.44).

(*j*) In a pond or stream study the way animals move. The main types of locomotion can be classified under the headings: amoeboid movement; movement by cilia; flagellation; creeping locomotion; swimming by legs; swimming by fins. Collect specimens of each type (see Methods 2(*a*), p.27). Make notes on their locomotion, illustrating your account by drawings and diagrams to explain how they move.

(*k*) In the same way as in (*j*) study the different sorts of breathing in pond animals. Headings could be as follows: by diffusion through the body surface; by gills; those which set up special water currents to make the water pass through their gills; those which breathe by coming up to the surface for an air bubble; those which breathe air by means of tubes; those which have lungs.

(2) A STUDY OF ONE PARTICULAR KIND OF PLANT OR ANIMAL

After a special habitat has been studied for some time by the whole class, each group can select one of the plants or animals there for further and more detailed study. This work is best done by two or three pupils working together and the aim is to find out as much as possible about the plant or animal. Sometimes it is more convenient to study a small group of animals rather than one special type. For instance, we can study 'mayflies' rather than one particular species of mayfly, because although there are several different sorts of mayfly, they are all more or less alike in their appearance, life history, and habits.

Here are some ideas to choose from: a butterfly or moth; frogs, toads or newts; mayflies; dragonflies; gnats; water fleas; greenfly; the earthworm; a bird, such as the robin, house-martin, swallow, blackbird, tits, thrushes and many others; a small mammal, such as the field mouse, vole or shrew; leeches; pond snails; land snails; a tree such as oak, ash, horse chestnut, beech, elm, hazel or Scots pine; a flowering plant, such as buttercup, dandelion, shepherd's purse, stinging nettle, groundsel or very many others; a parasitic plant, such as dodder, yellow-rattle, red rattle, eyebright and toothwort (some of these are not completely parasitic); a parasitic animal, such as the larva which feeds on cabbage white butterfly larvae.

Some ways of studying your chosen plant or animal are given below.

(a) Find out its name (see Methods 11, p. 56).

(b) Where does it live? Draw a map of your district showing where the animal or plant is to be found (see Methods 1, p. 26): try to search the district thoroughly so that your map is complete. Other maps or diagrams may be useful too – for example, quadrats (see Methods 6(a), p. 33) or profiles (see Methods 6(b), p. 44). Can you think of any reasons why it lives there and why it does not live in any other place? If not, wait until you have made some of the other studies described below.

(*c*) What is its life history? Study your plant or animal throughout the year. As far as possible you should study it in its natural surroundings since it may be affected if it is brought into the laboratory and will not grow or behave naturally. Make notes, illustrated by drawings and diagrams, to cover all the following points and as many others as you can think of.

Its eggs or seeds: What are they like? Where are they laid? How long do they take to hatch or germinate?

The young stages: Is it a small version of the adult or are there one or more young stages which are quite unlike the adult? Describe all the young stages, including how they feed, what they feed on, their locomotion, their habits, their method of breathing.

The adult: What is it like? Describe its feeding methods, food, method of breathing, locomotion, habits (including mating, nest-building, etc.), responses to stimuli (especially food, light, gravity and water), hibernation and migration (if any). Describe its method(s) of reproduction and give an estimate of the number of young produced (or number of seeds per plant). How do the eggs, seeds or young spread to new areas? In the case of a tree, describe the changes taking place during the course of the seasons, in addition to the growth of the tree from its seed.

(*d*) Study its food and feeding in more detail. Make a complete list of all the food it will eat. (This does not apply to a plant.) You can discover a lot about feeding by watching the animal in its natural surroundings. Also, you can try to capture some animals and supply them with different foods to see if they eat them. If you are studying a small mammal it is interesting to find the weight of the food eaten each day and compare this with the weight of the animal.

Another way of finding out about food is to examine the digestive system of the animal. With many small ones, such as water fleas and mayfly larvae, the digestive system can be examined by putting the animal under the microscope, for it is transparent. With larger animals it will be necessary to kill them and dissect out the digestive system. You can also examine their droppings (birds' droppings often contain seeds), the places where they feed (for uneaten pieces), and, with the owl and kestrel, their pellets. Can the animal get its food anywhere? If not, is the animal found in a particular

13

place because that happens to be the only place where its food is?

(*e*) What feeds on the animal or plant? If your example is an animal, look for its enemies. Most animals are fed on by larger animals, but occasionally an animal will be fed on by smaller ones, especially if it is dead. If your example is a plant, look for animals which eat the whole plant, those which nibble at the leaves and stems, those which eat the root, and those which tunnel into the plant, eating as they go.

(*f*) Taking the results of (*d*) and (*e*) together, build up a food chain including your example.

(*g*) What other plants and animals does your example need for food, shelter, nest-building, and dispersing its seeds or eggs? Does one of these requirements restrict your example to one particular habitat? For example, a bird that nests in trees must live where there are trees — in a wood, a park, or perhaps a garden.

(*h*) If it lives in water, has it any special features which make it particularly suited to living there?

(*i*) Is it of any importance to man? Is it useful? If so, in what way? (It might destroy pests, pollinate fruit trees, etc.) Is it harmful? If so, in what way? (It might be one that destroys man's food, his clothes, or his houses, or one that carries disease to him.)

(3) SPECIAL STUDIES

Here are several suggestions for projects that can be carried out by small groups. Some of them require special habitats (for example, heath or seashore) which may not be within reach of some schools but they can often be adapted to suit local conditions.

(*a*) Study the plant and animal population of a rotting tree stump or log. Look for animals in the bark and wood and look for fungi growing out of the tree. What do they feed on — the tree stump or each other? This project will mean a lot of work with a hand-lens and a microscope.

(*b*) Stake out an area of ground one metre square. The corners should be marked by four wooden stakes about 30 mm x 30 mm x 0·3 m long. The stakes should be painted or creosoted to prevent them from rotting. Drive them in with a mallet until their tops are about 10 mm above the soil. With a spade or fork dig the square thoroughly to a depth of at least 0·3 m. Remove every piece of vegetation, taking great care to take out the roots from the soil. The vegetation and soil outside the square should be left as undisturbed as possible. Level off the soil. By means of a map (see Methods 1, p. 26) record the position of the square so that you can find it again.

Return to the square once every month or two. At each visit plot a diagram to show what plants are growing on the square. (Treat the square as an ordinary quadrat according to Methods 6(a), p. 33). You should do this (or arrange for someone else to take over this project if you cannot continue) for at least two years. Put all the diagrams away carefully, as they are made. You will then have a complete record of how a bare piece of ground is colonized by plants. Which plants are the first to appear? How did they get there? Which plants come next? How do they get there? Do the first arrivals give way to the later arrivals? How long is it before the vegetation on the square is the same as that outside the square — or does it never return to its former state?

(c) This is similar to (b) but in this case the vegetation is removed by burning. Select an area of heath as soon as possible after there has been a heath fire. Near the edge of a burnt patch stake out a quadrat as before. Record the plants immediately (if there are any) and every month or two following. Make the same observations as for (b) above, but notice that in this case only the tops of the plants have been burnt off and that roots may be alive a few centimetres deep in the soil.

(d) On the seashore, some plants and animals are to be found only where the highest tide reaches. They cannot exist if they are submerged in the water for long. Others live only just below the lowest tide level and they must always be covered by water if they are to live. In between these levels will be found other plants and animals which can stand immersion in water or exposure to the atmosphere to greater or lesser extent. The result is that the plants (mainly sea-weeds) and many of the animals live in zones at different levels along the seashore. By means of transects and profiles (see Methods 6(b), p. 37), study the zones on a convenient part of the seashore. Can you discover any features which make a particular plant or animal suited to the zone it is found in? It is best to consider the following zones (see Fig. 1):

A. Just above the high spring tide level — salt spray only.

B. Between high neap tide and high spring tide levels — covered only at high tide and not covered at all on some days.

C. Between low neap tide and high neap tide levels — covered and uncovered twice daily.

D. Between low spring tide and low neap tide levels — uncovered only at low tide and not uncovered on some days.

E. Below low spring tide level — always covered.

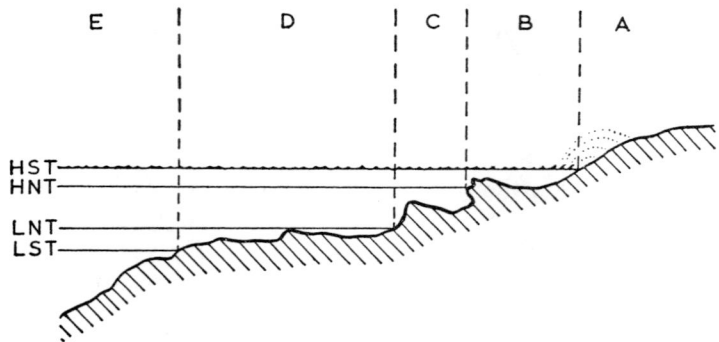

FIG. 1. Tide-levels and zones on the seashore. HST = high spring tide. HNT = high neap tide. LNT = low neap tide. LST = low spring tide. Letters A—E refer to zones listed above.

(*e*) Investigate the effects of trampling on vegetation. This can be done by studying a footpath across a heath or a field or in a wood. Use transects or profiles (see Methods 6(*b*), p. 37 and Fig. 9) to illustrate your results.

(*f*) Make a study of insect pests in a garden or a field. Collect as many specimens as you can from different plants, identify them and write an illustrated account of these pests and the damage they do.

(*g*) Study climbing plants in hedgerows (or any other place where you can find a worth-while number of climbers). Collect specimens, press them (see Methods 5, p. 32), and sort your collection into different groups according to the method which the plant uses for climbing.

(*h*) Set up a bird table or a nesting box in a secluded part of the school grounds. Keep careful, dated records of the birds seen at the table, how many there are at one time, and their behaviour, feeding habits and reactions to other kinds of bird. If possible, watch the bird table at different times of day and find out if different kinds of bird come at different times.

Try putting out dishes containing different sorts of food, such as raisins, peanuts or peanut butter, bread, suet or bacon fat, sunflower seeds, maize seed and millet. Note the food preferences of the different kinds of birds.

(i) Make a study of insect-eating birds, such as the swallow or house-martin. Find a nest in which the eggs have just hatched. From a distance, watch the nest and count the number of visits made by the parents during the course of 1 hour. Repeat this on several successive days, at different times of day if possible. From your figures you will be able to estimate how many insects are required to feed each young bird up to the time it leaves the nest.

(j) Make a study of weeds found in a garden or a field (including the school playing-field). What are the names of the weeds? How did they get there? What damage do they do? Are they easy to get rid of? If they are not easy to get rid of, why not?

(k) Remove some of the surface soil from a garden bed or other fairly open ground. Remove all plants, including roots, from the soil. Place the soil in a shallow wooden box in a greenhouse or indoors by a window. Keep it watered and watch for seedlings. Let them grow to the flowering stage and find out their names.

(l) Find out the effects of cultivation on garden weeds. Clear a garden bed of all plants and then divide it into a number of plots. Cultivate each plot differently: leave one alone; dig another thoroughly, turning over the top 0·2 m of soil; hoe another plot once; hoe another plot every month; scrape the top 1 cm of another plot every month. Record the number of weeds of different sorts which grow on each plot. Which is the most effective way of keeping the soil clear of weeds?

(m) Grow some vegetable or garden flower plants in a garden bed. Let one half of the bed become very weedy but carefully remove all weeds from the other half of the bed at least once a week. What effect do the weeds have on the growth of the vegetable or garden plants? Record the effects on their height, their colour, their flowering, the number of fruits or seeds, and their dry weight at the end of the experiment.

(*n*) Study the reproduction of birds to see how successfully birds propagate. Look for nests in which there are eggs. Every few days return to the nest and count the number of eggs. In due course count the number of nestlings which have hatched and later on record the number of young birds which have grown large enough to leave the nest and lead an independent life. You must record the date of each visit with the other details. If possible, study several nests of the same sort of bird. What is the average number of eggs laid by each hen? What percentage of these hatch? If any fail to hatch, why? What percentage of the nestlings develop sufficiently . well to be able to leave the nest? What enemies have the young birds?

(*o*) Dig a trench in any habitat you are studying. Draw a map of one vertical side of the trench to show the roots of the plants. Your map should show the depth at which each plant forms most of its roots and how far sideways the roots spread. In a given habitat you will find that different sorts of plants root at different depths — what advantage is this?

(*p*) On a warm, sunny day in summer spend an hour or more watching flowers in a garden, a wood, or by the roadside. Notice which insects visit each sort of flower and record your observations. Collect a specimen of each sort of insect for accurate identification.

(*q*) Sow spinach seeds in two plots. One plot must have soil which contains no humus and the other plot must have soil with humus and fertilizer added to it. When the spinach plants are fully grown, notice any differences between them. Is one lot of spinach more subject to insect pests than the other? Can you suggest any reasons for this?

(*r*) In the autumn, plot a belt transect 0·5 m wide between the bases of two trees (see Methods 6(*b*), p. 37). On the transect plot the positions of all fallen seeds or fruits. How far have these seeds or fruits travelled from the parent plant? Does the direction of the prevailing wind affect their distribution?

(*s*) Plot quadrats (see Methods 6(*a*), p. 33) to show the plants growing on a lawn or playing-field (*a*) near the middle, where it is well mowed, and (*b*) at the edge where the grass is not mowed. What are the differences? Explain them if you can.

(*t*) Find a tree with *Protococcus* growing on it. Notice that it only grows on certain parts of the bark. Why is this? In trying to discover the answer, make observations of the tree and other trees and try some experiments of your own devising.

(*u*) Try growing fungi and bacteria from different soils. You can compare garden soil, soil which has been manured or to which chemical fertilizers have been added, soil from different districts and so on. Weigh out 50 g lots of each soil into petri dishes. To each lot of soil add 1 g powdered calcium carbonate, 2 g starch and 0·5 g glucose. Add water and stir to make a creamy paste. Put the lids on the dishes and leave them in a dark place. Look at them every two or three days and see what kinds of fungi or bacteria grow. Which soils contain the most fungi and bacteria? Which contain the least?

(*v*) Collect eight frogs, noting the colour of the mud or other background on which they were resting when you caught them. Are all the frogs nearly the same colour? Put four frogs in each of two small glass tanks. Stand one tank in a shallow box which is dead black inside. Stand the other in a shallow box which is white inside (or you can stand it in a white sink, or enamel dish). After about 2 hours report on the colour of the frogs in both the tanks. Return the frogs to the place from which you got them. Do they match their background as well now as they did when you caught them? You can try similar experiments with other animals, such as tadpoles, sticklebacks, pond skaters, larvae of the large white butterfly and other animals which you can catch. Butterfly larvae should be kept in light or dark surroundings and left to pupate. Contrast the colour of pupae developing in these two conditions.

(*w*) Collect about fifty insects from each of the following situations:

(i)   A grassy field or lawn.

(ii)   The litter below a hedge or below trees.

(iii)   Wooden fences, walls of wooden buildings, and the bark of trees (you may find few insects here, so be content if you find only twenty).

(iv)   Mud beside a pond or river.

(v)   Dense clumps of all one kind of flower.

Sort each set of insects into groups according to their main colour. The following groups should be distinguished:

(i) Black.
(ii) Grey or brown.
(iii) Buff.
(iv) Green or blue.
(v) Red, orange, yellow or white.

If any insect clearly shows two or more colours, you should count it under two or more of the above colour groups.

Make a table showing how many insects of each colour group were found in each situation. Where were most black insects found? Is this an area in which the main background colour is black? Or are there many black shadows in this situation? Where were most of the brightly-coloured insects found? Can you suggest why this was so? Similarly try to explain the distribution of insects of the other colour groups. Remember that many insects rely on camouflage to help them escape from enemies, while others have bright warning colours to scare their enemies away. Often the bright insects have an unpleasant taste or a powerful sting.

(x) Prepare five shallow trays about 20 cm square and 2 cm deep. The trays should be coated inside as follows:

(i) White rough surface — mix some white paint with coarse sand and fine gravel. Paint this on to the inside of the tray. If you wish to keep the tray clean for other purposes, cut a square of plywood to fit the bottom of the tray and paint this.

(ii) Black rough surface — as above, but use black paint.

(iii) Blue or green rough surface — as above but use a medium shade of blue or green.

(iv) Soil or gravel surface. Coat the tray with a film of glue, sprinkle soil or gravel or a mixture of both on to the glue. When dry tip off the surplus.

(v) Mottled rough surface. Paint the tray with small spots of the paint or soil mixture used above.

Set out the trays on a lawn or some other open space. The trays should be close together but not touching (Fig. 2). Preferably the area chosen for this investigation should be generally undisturbed by passers-by. Sprinkle bird-seed around the area, both on the trays and on the ground between and around the trays. You can use a mixture of several kinds of seed, sold in pet shops as 'Wild Bird Seed'.

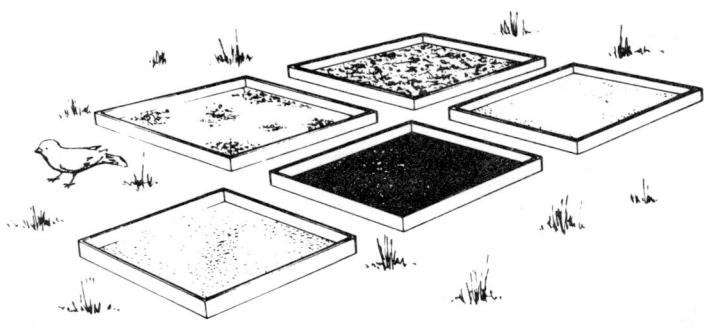

FIG. 2. Setting out the feeding trays for Special Study (x)

Repeat this for two or three days; this will accustom the local birds to feeding in this area and they will become familiar with the sight of the trays. On the third or fourth day empty the trays. For each tray count out seeds to make a mixture of known composition. As far as possible your mixture should include seeds of several colours. For example, you might count out: five large red seeds (peanuts), five large black seeds (sunflower), ten large buff seeds (split peas), fifteen medium-size brown seeds (barley), thirty small black seeds and thirty small white seeds.

Sprinkle this mixture on each tray, and sprinkle some of the usual mixture on the ground between and around the trays. Leave the trays until the birds have been feeding for a while, but not so long that they will have had time to eat the majority of the seeds. Do not leave the trays overnight, for mice may come, which will upset the results of the investigation. This is an investigation of the effect of background *colour* on the feeding of *birds*; *mice* feeding in the *dark* would show quite different results.

When the birds have fed for long enough (you may need a trial or two to determine the best length of time) tip the seeds from each tray into a white dish. Count how many seeds of each kind are left, and in this way find out how many of each kind have been eaten. Since you put different numbers of each kind in the mixture you should express the number of seeds eaten as a percentage of the number of seeds supplied. Were more black seeds taken when they were on a black background or when they were on a white background? On which background were least buff seeds taken? Were some

21

kinds of seeds almost completely taken on any kind of background? Study your results in this way and try to explain any
differences you notice. It may be that the colour of background makes it hard for the birds to discover certain
kinds of seeds. Also some birds may show preferences for
some kinds of seed, and leave other kinds of seed even though
they can see them plainly. The results of your investigation
may give you some ideas for further investigations. For
example, if birds seem to prefer one kind of seed you could
count the seeds on the trays after a short time, then return
them to the trays, leave them longer and count again later.
You could then tell if birds preferred one kind of seed, but
would eat other kinds after the preferred kind have been
eaten.

(4) MAKING COLLECTIONS
There are three good reasons for collecting:

(*a*) In order to take back specimens so that they may be
identified or so that their structure or habits may be studied
more closely than is possible in the field.

(*b*) So that you may have specimens to put on display at
school for use in connexion with lessons.

(*c*) To make a reference collection. A collection of all the
common plants from, say, a local wood will help those who
are doing field work there.

There are also several *bad* reasons for collecting; one of
these is that *collecting is fun.* So it is, but you should never
collect plants and animals simply for fun. There are many
boys and girls who love to spend hours at a pond collecting
the animals. They bring them home, put them in a tank or a
jam-jar and then *forget all about them.* Unless you definitely
need to bring the plants or animals back for one of the three
good reasons you should leave them where they are – in their
natural surroundings. *If* you have a good reason for collecting
you can have the fun too, but it is selfish to have your fun at
the expense of wild life.

Some pupils make the mistake of collecting too many
animals – which is nearly as bad. They intend to study them
all when they get them home but forget that time is limited
and so a few get studied and the rest are left to die. So do not
collect too many. Remember, it is easy and quick to collect a
lot of specimens but it always takes far longer to identify them

or study them properly in the laboratory. If there are lots of different animals at the place you are visiting, collect a few of them (no more than you will actually have time to deal with when you get back to school) and make a note to come back again some day for a few more. When you have finished with your specimens return them to their natural surroundings as soon as possible.

There are things which you should *never* collect. For instance, do not take plants or animals that are rare or uncommon in the place you are visiting. There is so much to find out about even the commonest ones that there is no need to collect rarities. Thoughtless and selfish collecting can easily cause rare plants and animals to become extinct in a short time. Finally do *not* collect birds' eggs, not even for a reference collection. You can easily get a book containing colour photographs of eggs and this will be just as good as a reference collection — better, in fact, for it will slip into your pocket and it will not break if you drop it.

Some methods of collecting, keeping, and preserving specimens are given in Methods 2—5.

Here are some ideas for collections:

Collections for display in the laboratory:

(i)   Fruits and seeds, illustrating various methods of dispersal from the parent plant (you will get most specimens in autumn).

(ii) Large fungi. Best collected in early autumn, especially after a period of warm, muggy weather. Throw the specimens away when finished with.

(iii) Climbing plants.

(iv) Living specimens of the animals from one particular habitat. Label each specimen with its name and where it is to be found. As far as possible show each specimen in its natural surroundings (water fleas in a tank of water, wood-boring grubs in a glass-covered case containing wood, and so on).

(v)   Pressed specimens of one sort of plant at different stages in its life: a seed; a seedling; a young plant with leaves just unfolding; when the flower buds are forming; when it is in flower; when the petals are dropping off and seeds are forming; the plant at the end of its life (if an annual) or in its over-wintering form (if a biennial or perennial).

(vi) Moths to show wing-markings for camouflage. In each

case record the background from which the moth was taken (its colour, light or dark, rough or smooth, uniform in colour, or mottled). If possible, collect two specimens and display one on a white card and the other on a piece of the bark or other natural background on which it was found.

(vii) Land snails with striped shells in various colours. These colours are a camouflage and, as in the previous collection, you should record the background and display your specimens on these backgrounds, as well as on white card.

Collections for reference:

(i) Plants, pressed. It is better to sort them into folders according to their habitat. Have folders entitled Woodland Plants, Hedgerow Plants, Seashore Plants, and so on.

Some sorts of plant (for example, shepherd's purse, stinging nettle) will be included in more than one folder because they are found in more than one habitat.

(ii) Beetles.

(iii) Butterflies and moths.

(iv) Cones from coniferous trees.

(v) Mosses and liverworts — these can be dried and kept in match-boxes, pill-boxes, or specimen tubes.

(vi) The bark and leaves of trees.

You will be able to think of other collections which will be of use to you in your field work.

METHODS

In this section you will find detailed instructions to help you do your field work properly. The methods are ones which any intelligent boy or girl should be able to carry out on his or her own. When you have read through the previous section and decided what sort of field work you wish to do, look through this section to find out how to set about it.

The methods are here for you to use — *do not let the methods use you.* Methods are not field work and no amount of map-drawing and quadrat-making will help you find out about the lives of plants and animals unless you remember always that maps, diagrams, and lists are only a *means* to an end — they are not an end in themselves. The aim of field work is to study living plants and animals in their natural surroundings. A lot of useful work can be done with no special methods and no apparatus — just a sharp mind and a note-book. But some people will want more precise measurements, charts, maps, and diagrams to help them in their work; these are the people who will need to *use the methods* which follow.

(1) MAPS

It is often important to have a map of the area you are studying. It will illustrate the main features of the area and it will help you to locate special parts of the area when you visit it again. If the area is a large one, it will be marked on an Ordnance Survey map. Tracings taken from a map can form the basis of your own maps. Trace the outlines and then fill in the details you require, by sketching or by measuring with a surveyor's tape.

The simplest method is illustrated in Fig. 3. Suppose that you wish to plot the position of a rabbit's burrow in an area of heathland. Trace the outline of the heath (Fig. 3A). Stand by the burrow and notice that it lies on a line which joins the two corners of the heath, *b* and *c*; also it lies on a line joining corner *a* with a corner of the barn. Draw these two lines lightly on your tracing (Fig. 3B). The place where

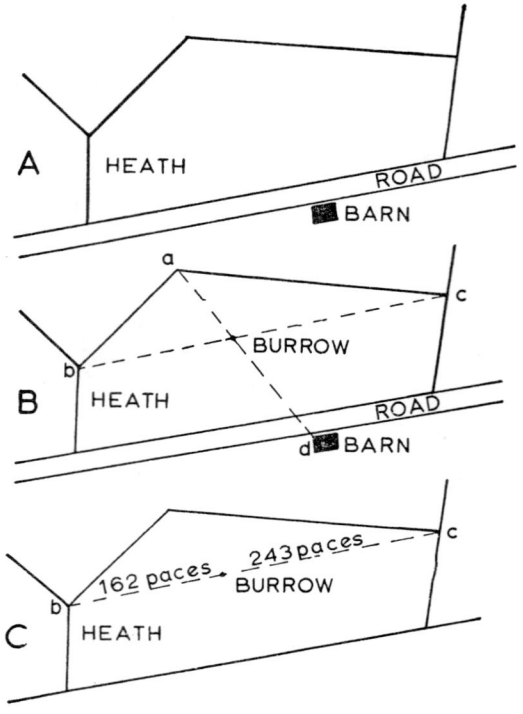

FIG. 3. Simple ways of mapping

they cross will be the position of the burrow. Another simple way of doing this, shown in Fig. 3C might be useful if your burrow happens not to lie on two easily plotted lines. You notice that the burrow lies on the line joining b and c, as before, but you are not able to find another line (as there is no barn in this diagram). Walk from b to the rabbit burrow, counting how many steps you take; then walk from the burrow to c, counting again. In this example the distances are 162 paces and 243 paces, a total of 405 paces. Measure the distance from b to c on the map; suppose in this example that this distance is 10 cm. Now you can reason as follows:

thus

10 cm on the map represents 405 paces on the heath,

$\frac{10}{405}$ cm on the map represents 1 pace on the heath,

26

therefore

$$\frac{10 \times 162}{405} \text{ cm on the map represents 162 paces on the heath,}$$

$$\frac{10 \times 162}{405} = 4 \text{ cm.}$$

Draw the line from *b* to *c* and measure 4 cm along the line from *c*. This point marks the position of the burrow.

The sort of method described above is not very accurate but it is quick and often good enough for the sort of map you will need. You will realize that measuring distances by pacing is not very accurate, especially on rough ground like a heath, because it is not easy to take paces which are all equal in length. A more accurate yet simple way of measuring distances over fairly level country is a trundle wheel which consists of a large wooden wheel on a stick. The trundle wheel, made by E.J. Arnold and Son Ltd., has a circumference of exactly one metre. The wheel is placed on the ground and as it is pushed along the distance to be measured the wheel rotates. Every time it goes round once means that a distance of 1 metre has been covered. The wheel has a clicker which clicks once for each turn, so as you push it along you have only to count the clicks and this indicates the length in metres. This is a much more accurate way of measuring distances than pacing, but you must take care not to let the wheel slip. Over rough ground, when a trundle wheel will give too large a figure, a more accurate method is to use a surveyor's chain or tape.

(2) COLLECTING
(a) Animals.
(i) *Simple methods.* Dig in the soil with a spade or trowel; prise pieces of bark from logs with a stout knife or chisel (but do not take off large pieces of bark from healthy trees — they will soon become unhealthy if you do); look under stones; take water from a pond or stream (for microscopic animals and plants); take the soft mud from the bottom of a pond or stream.

(ii) *Baiting.* Attach pieces of potato, carrot or swede to a skewer or a small stick. Bury the bait in the soil with the skewer sticking out so that you can find it again. Dig it up later. Millipedes, wood-lice, and other small animals will be attracted to the bait.

Place some raw meat in a small glass jar on the bottom of a pond or slow-moving stream. This is the way to catch free-living flatworms.

(iii) *Nets.* There are various types of net for catching different sorts of animals. The butterfly net has a wide mouth and a short handle and is made of soft, fine fabric. A sweeping net is more robust and is used for sweeping through grass and other tall vegetation to catch insects which are hidden there. A fish net is on a long handle (often in sections that can be screwed together) and is made of a coarse fabric. A plankton net is made of very fine fabric and has a glass or plastic tube at the bottom; microscopic animals collect in the glass tube when the net is swept through the water (take care not to break the tube when using this net). A pond drag is a net mounted on a frame that can be pulled along the bottom of a pond or stream on the end of a long rope.

Any net that has been in water or has become wet in use must be cleaned, washed in fresh water and left out to dry as soon as you return to the laboratory.

(iv) *Beating tray.* This is a framework on which is stretched a dark-coloured cloth. It is held or placed underneath a tree and the branch of the tree is shaken or banged with a stout stick. This dislodges small insects and other animals which are on the branch and leaves. They fall off and are caught on the tray. Instead of a proper beating tray you can use a cloth held beneath the tree or spread out on the ground. Another idea is to hang an open umbrella on the branch by its handle.

(v) *Traps.* The simplest trap is a jam-jar sunk into the ground with its mouth level with the surface. Place some bait (meat, bread, fruit) inside the trap and come back on the following day. To collect small mammals such as voles, field mice, and shrews, use a Longworth Trap. Nesting material (wood shavings, straw) and bait are put in the trap and the door is set. The animal enters the trap and the door closes. Small mammals need enormous amounts of food so never leave the trap for longer than 12 hours without visiting it.

(vi) *Collecting earthworms.* Use a 40 per cent solution of

formaldehyde or a deep red solution of potassium perman-
ganate. Pour the solution on to the soil. After a while the
worms will come out of their burrows. Rinse them in water
immediately or they will die.

(vii) *Small animals in leaf litter or soil.* The cheapest way of
making this apparatus is to use a large plastic funnel about
15 cm in diameter (from any ironmongers). Cut off the stem
of the funnel and fix a disk of wire gauze or perforated metal
inside the funnel, about 3 cm from the top. The holes in the
gauze or metal should be about 5 mm across. Place the
sample of leaf litter or soil on the gauze, spreading it evenly
to cover the gauze. Arrange a bench lamp, with a 40-watt or
60-watt bulb, to shine down on the litter and warm it. The
surface layer of soil should not get hotter than 40°C — check
this with a thermometer from time to time and adjust the
position of the lamp accordingly. The warmth and light will
cause the small soil animals to move downward through the
layer of litter, through the gauze and down the funnel.
Underneath the funnel place a small beaker of water to
collect the animals as they fall out. This takes several hours.

(b) Plants.

Some of the smaller plants, the one-celled algae and the
like, can be collected with a pipette or a plankton net. No
special apparatus is needed for collecting larger plants,
though a trowel is useful. Collect whole plants whenever
possible: roots, stems, leaves and flowers. Collect some
ripe seeds and fruit too, if these are available. *Never*
collect rare plants.

(3) CARRYING SPECIMENS HOME

Always take a good supply of containers — more than you
think you will need. You will then be ready for the unexpec-
ted specimen you may be lucky enough to find.

Before you set out, put a blank label on each container so
that you will be able to enter details of the specimens you put
inside.

(a) *Jars and specimen tubes.* Wide-mouthed jars with perforated screw-on lids can be used for larger animals. Smaller
animals can be put in glass specimen tubes with cork stoppers
or in plastic ones which usually have a plastic stopper attached.
Some of the jars and tubes should have plaster-of-Paris (about

2 cm deep) at the bottom. Moisten this with water to keep the air damp in the jar. Small animals will live for a long time in these jars.

Do not put too many animals in one jar — they may eat one another before you get home. Therefore take plenty of jars.
(b) *Boxes.* Pill-boxes (with glass lid or with cardboard one), match-boxes and plastic sandwich-boxes are useful for many sorts of specimen.
(c) *Vasculum.* This is a large, flat tin for carrying plants. It usually has a shoulder strap. If you have no vasculum, use polythene food-bags or a large air-tight tin. Plants become dry very quickly so keep the vasculum shut whenever possible and do not leave it in hot sunshine. If you have only a few plants in it put some large leaves in the vasculum to keep the air damp. Plants can be labelled as they are put into the vasculum by fixing small tie-on tags to their stems. These are normally used in shops as price-tags and they can be obtained from a stationer's. Alternatively, record details in your field note-book. Make an entry like this: 'Plant with large blue flowers and grey-green, very hairy leaves — found in long grass by roadside.' As long as there is only one plant in your vasculum which fits this description, you will be able to sort it out when you get home and then find its proper name.

(4) KEEPING PLANTS AND ANIMALS

Animals and plants which you have collected in the field may often be more conveniently studied in the laboratory or greenhouse, provided you remember that they may not behave in the same way in captivity as they do in their natural surroundings. As far as possible you should try to set up an exact copy of the type of habitat in which the plant or animal was found. Here are some ways of keeping plants and animals.
(a) *An aquarium.* This should be a large, fairly shallow vessel. Stand it where it will not receive the direct light of the sun. Place some washed gravel or sand in it, and put a piece of paper on top of the sand. Pour water on to the paper; the paper will prevent the sand from being washed away and the water will not become cloudy with suspended particles from the sand. Obtain some healthy water plants, tie their stems to stones, and sink them near the back of the aquarium. Allow the aquarium to stand for a week before introducing animals.

At the end of the week, siphon off the water and replace it with fresh water. You will need some sort of apparatus for bubbling air through the water; an electric air-pump is best. Aerate the water for a few hours before putting in animals, if you are using tap water in the aquarium; this will drive off dissolved chlorine gas which would be poisonous to the animals. While waiting to transfer your animals to the aquarium, bring them into the same room, still in the jars of pond-water in which they were caught; this will allow the pond-water in the jars to come to the same temperature as the water in the aquarium. You may then pour your catch into the aquarium.

In stocking your aquarium, bear the following points in mind. Avoid overcrowding. Remember that some animals will attack other ones. Keep the tank clean, removing all dead animals and plants and any uneaten food. Pond snails are useful because they will eat the thin green film of algae that grows over the glass and makes it difficult to see into the tank. Change the water whenever it becomes cloudy: remove the animals, and then siphon off the used water and replace it with fresh, aerated water that is at the same temperature as the used. Replace the plants from time to time if they are eaten away by any of the animals; plants are important in an aquarium because they provide shelter for the animals and food for some.

(*b*) *Hay infusion.* This is a 'habitat' in which Protozoa will live. It is interesting to set up a hay infusion and examine its population each week. The first animals to develop may later disappear and be replaced by others. This is a piece of 'field work' that can be done indoors in the winter.

(*c*) *Earthworms.* These and some other soil animals may be kept in soil in a large box which has one or more glass sides. The soil should be kept watered and there should be drainage holes at the bottom so that it does not become waterlogged. Cover the glass with a sheet of thick card or wood to keep out the light, except when you wish to observe the animals. They will then burrow close to the glass. A few leaves placed on the surface of the soil will provide food for the worms.

To prevent the worms escaping, the top of the box and the drainage holes should be covered with fine wire gauze.

(*d*) *Insects, amphibia, reptiles, birds and mammals.* There are many good books which deal in detail with the keeping of these animals. You should refer to some of those listed in the Book Lists, pp. 59—60.

(*e*) *Plants*. The simplest method is to cut a turf, about 50 cm square and 15 cm thick, from the soil where the plants are growing. Place it in a shallow box in a greenhouse or by a window. By this means, samples of the vegetation from various habitats may be studied side by side.

Different habitats may be set up in the school garden and planted with specimens gathered 'in the field.' An old sink or bath can be filled with soil and made into a 'swamp' or 'bog'. A rockery which is well drained provides a good place for growing plants from dry habitats. The type of stone used for the rockery is important: for instance, plants from limestone areas will usually only grow well in a limestone rockery. It is not possible to go into full details here; the essential thing is to study the natural habitat in detail and then try to reproduce all its essential features in the school garden.

(5) PRESERVING SPECIMENS

You will do this only if you need specimens for a reference collection. All other specimens should be released in a suitable place if they are alive. There are many simple books on insects amongst those listed on pp. 59-60. Plants should be pressed according to the method described below:

(*a*) Wash soil off the roots and blot them dry.

(*b*) Spread the plant between folds of absorbent paper. It is possible to buy special drying paper, but newspaper will do. Try to arrange the plant so that its leaves and stems are in a natural position. Make sure that the leaves and flowers are flattened, not wrinkled and bent. Make a pile of folded sheets, each containing different plants.

(*c*) Place the pile of sheets under gentle pressure — use a plant press or failing this, a few heavy books. The plants should be pressed firmly but carefully, for too much pressure will squash them.

(*d*) After a day or two examine the plants and change the paper, replacing it with fresh, dry paper. Hang the old, damp paper up to dry. Press again.

(*e*) Repeat (*d*) every few days until the plants are flattened and dry.

(*f*) Fix them to a large sheet of white paper. You can use paste, small pieces of gummed paper, or small pieces of adhesive transparent tape. Put only one sort of plant on each sheet.

(*g*) On the sheet write the name of the plant, where it was found, the date, the habitat it was found in (*e.g.,* wood, rough grass, hedgerow), and the name or initials of the person who found it. You can buy printed labels which have spaces for all these particulars and look neat.

(*h*) Put the sheet into a folder with other sheets of plants from the same habitat.

Some plants have thick roots, bulbs, or very large, fleshy flowers that do not press flat very easily and take a long time to dry. Before drying these plants cut away the back of the thick part. This will not affect the appearance of the pressed specimen when seen from the front.

(6) DESCRIBING THE VEGETATION

There is only one way really to describe the vegetation of an area — draw a map which shows the position of each plant. But even a small wood with an area of only 1 hectare will contain perhaps 750 000 plants! It is obviously impracticable to count them, let alone map them. So instead of attempting the task of mapping them all we must be content with mapping only a few of them — a *sample* which is typical of the whole. The sample must be chosen so that it is representative of the whole or some particular part of the area which is under study. If there are two or three distinct types of vegetation in one habitat, samples must be taken from each of these areas. For example, in a wood the vegetation under the trees will probably be different from the vegetation in the more open parts of the wood, so a sample must be taken under the trees and another sample in the open. By comparing the two samples we can see the effect of shading by trees. There are several ways of taking samples and the one you choose will depend on what features of the vegetation you wish to illustrate.

(*a*) *Quadrats.* A quadrat is a map of a small area, usually square, and shows all the plants growing in that area. The size of the square varies; if you want to show the trees of a wood the quadrat might be 50 m square or more, to include sufficient trees; if you want to show how mosses and lichens are growing together on the surface of a rock, the quadrat need be only 10 cm square. For most purposes the size of the quadrat should be 1 m square. Here is the method:

(i)   Select the spot for your quadrat. It should be at a place typical of the area you are studying.

(ii) You will need four metal skewers and a piece of string just over 4 m long. Peg out the four corners of the quadrat with the skewers, wind the string round the skewers, and tie it firmly. This marks out the area of the quadrat (see Fig. 4A).

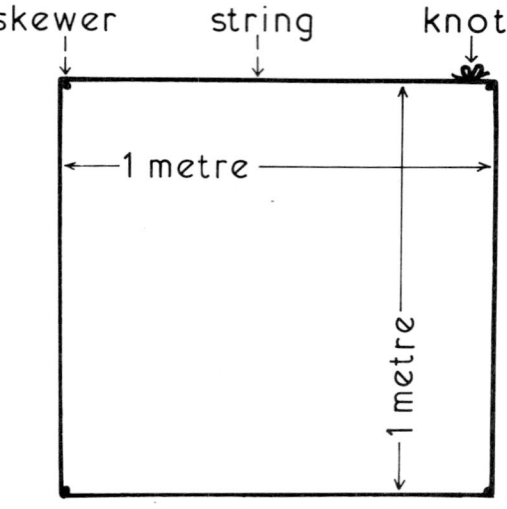

FIG. 4A. Plotting a quadrat

(iii) Now you will need eighteen pieces of string just over 1 m long, with a skewer tied to each end. Peg the skewers so that the strings stretch tightly across the quadrat and form a grid which divides the quadrat into 100 squares (see Fig. 4B).

(iv) On a piece of graph paper draw a square 10 cm × 10 cm. This will likewise contain 100 squares, each measuring 1 cm × 1 cm. Before you go any further, add a title or some other description which will help you to remember the location of the quadrat; a small sketch map might do. Also put some indication of direction, such as an arrow pointing north or to some prominent feature, the date, and the scale (see Fig. 4C).

(v) On your graph paper draw in the main features of the area — patches of bare earth, parts where one particular kind of plant is growing so thickly that you cannot distinguish the separate plants, large tussocks of grass, etc. The quadrat on the ground is divided into squares and that on the graph paper

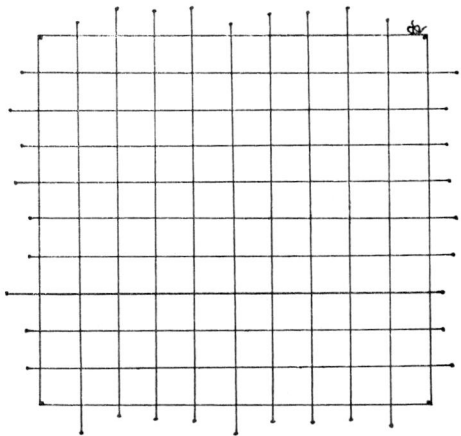

FIG. 4B. Plotting a quadrat (cont.)

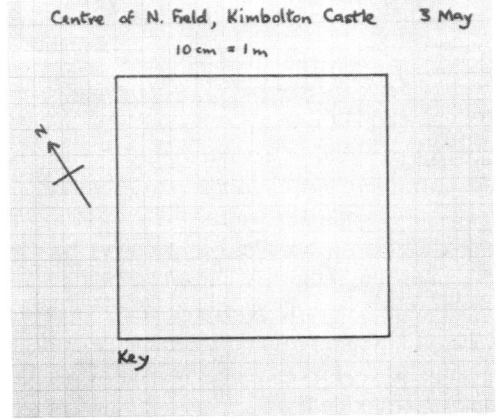

FIG. 4C. Plotting a quadrat (cont.)

is likewise divided, so it is easy to map the main features. As you map them, shade them or mark them with letters to indicate the different kinds of plant. Make a note of the meaning of all types of shading and letters (see Fig. 4D).

(vi) Finally, take each small square in turn and plot the positions of all the plants which have not already been plotted

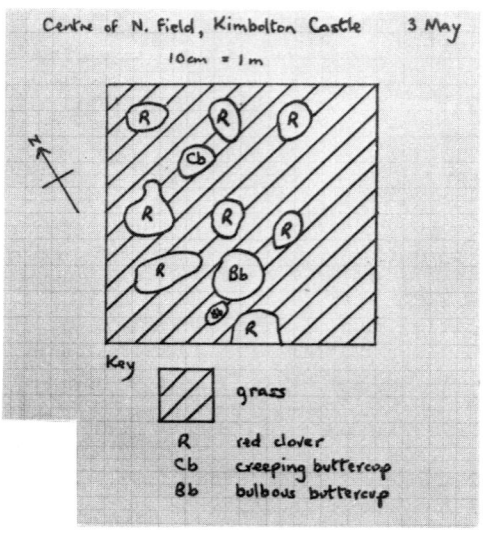

Centre of N. Field, Kimbolton Castle   3 May

10 cm = 1 m

Key

	grass
R	red clover
Cb	creeping buttercup
Bb	bulbous buttercup

FIG. 4D. Plotting a quadrat (cont.)

by shading (see Fig. 4E). This completes your field record.

When you come to write your results neatly, you may prefer to draw your quadrat in colour or use symbols. If you do this, it is easier to understand and easier to compare one quadrat with another. Fig. 5A shows a 'fair copy' of the quadrat from Fig. 4. Beside it, for comparison, is another quadrat made on the same day in a small wood, 150 m away. It is easy to see the differences between the vegetation of the field and that of the wood.

If you find it tedious to draw small circles and paint them, stick on small circles of coloured gummed paper instead. The circles are made by using a paper-punching machine or a leather-punch. Another idea is to use coloured gummed-paper shapes or the small self-adhesive circles sold for marking lantern-slides. If you have more different plants than you have colours, cut circles in half to make semi-circles or cut small squares from gummed paper. A yellow circle can be used to represent one sort of plant, a yellow semi-circle for

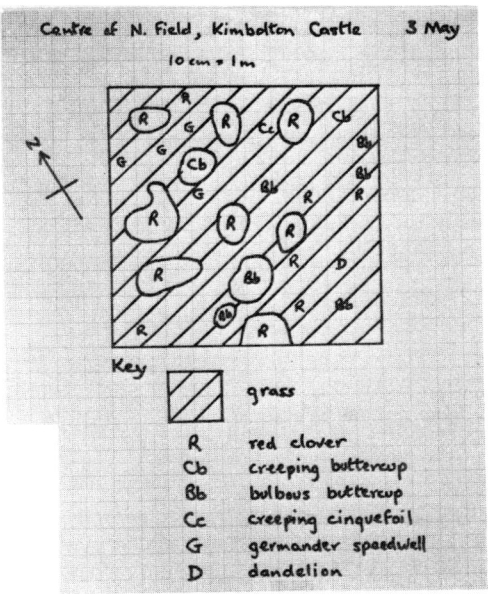

Centre of N. Field, Kimbolton Castle   3 May

10 cm = 1 m

Key

grass

R   red clover
Cb   creeping buttercup
Bb   bulbous buttercup
Cc   creeping cinquefoil
G   germander speedwell
D   dandelion

FIG. 4E. Plotting a quadrat (cont.)

another, a yellow square for another, and so on. Note that the grid lines have been left out of the 'fair copy' as they serve no useful purpose now.

Sometimes it is interesting to come back at another season and map the quadrat again. If you intend doing this, drive four stakes into the ground to mark the four corners of the quadrat before you leave (see also p. 34).

(b) *Transects.* These are cross-sections of the vegetation. A transect is a useful way of showing how the vegetation changes from one part of an area to another. The method is as follows:

(i) Decide upon the line along which your transect is to be made. If there is a distinct boundary between one sort of vegetation and another, the line should preferably cross this boundary at right-angles (see Fig. 6A). If the land slopes let the line run straight down the slope (see Fig. 6B). Mark the

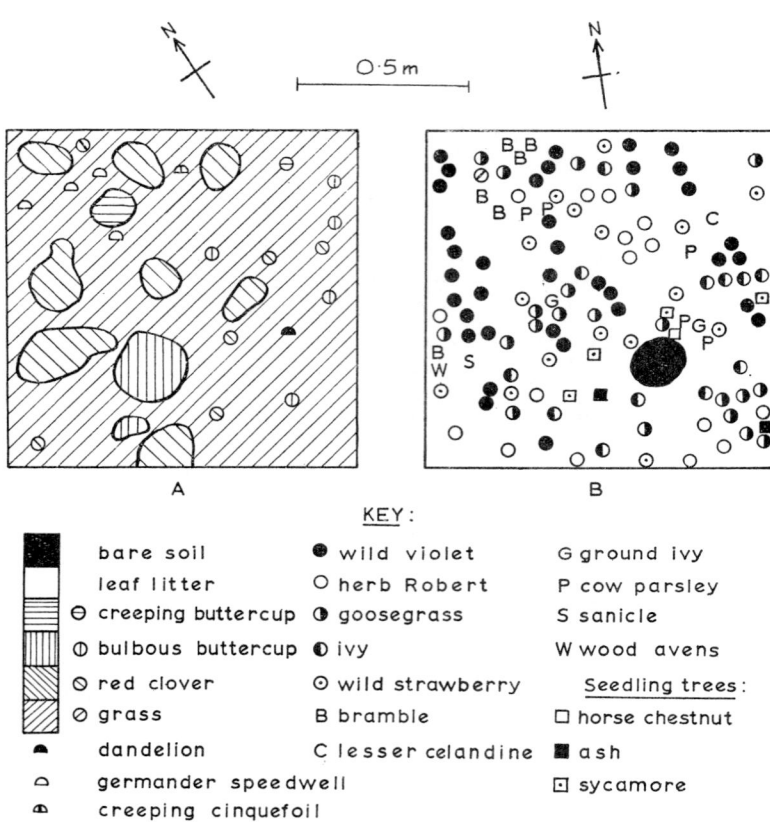

KEY:

■ bare soil	● wild violet	G ground ivy
leaf litter	O herb Robert	P cow parsley
⊖ creeping buttercup	◑ goosegrass	S sanicle
⦶ bulbous buttercup	◐ ivy	W wood avens
⊘ red clover	⊙ wild strawberry	Seedling trees:
⊘ grass	B bramble	□ horse chestnut
▲ dandelion	C lesser celandine	■ ash
◔ germander speedwell		▣ sycamore
◔ creeping cinquefoil		

FIG. 5. Two quadrats from different types of vegetation.
A: North field, Kimbolton Castle, Huntingdonshire.
B: Small wood near the gatehouse, Kimbolton Castle, Huntingdonshire.
Both quadrats were plotted on the same day in May.

These two diagrams represent two areas on the same day, about 150 m apart. In what ways do they differ? Notice that apart from one grass plant in the woodland quadrat (can you spot it?) these areas have no plants in common. There are many more different kinds of plant in the wood than there are in the field (fourteen kinds, compared with seven kinds in the field). In the field, the whole quadrat is covered by grass, with clumps of other plants and a few single plants here and there. In the wood there are no plants in clumps and there is plenty of bare earth covered by dead leaves showing between the plants.

Why are there differences between these two areas? Why are there no wild violets growing among the grass in the field? Why does grass grow in a continuous carpet, so that we cannot pick out the separate grass plants? Why is there no speedwell in the wood? Does speedwell ever grow in a ends of the line with stakes (bamboo garden canes will do).

(ii) In your field note-book record the location of the transect line, using sketch maps like those in Fig. 6, and add the date and any other items of interest.

(iii) Lay a tape measure along the line, from one stake to the other, with the end of the tape at one of the stakes. If your line is longer than your tape, divide the line into sections by putting other stakes at intervals along the line. Then work on each section in turn.

(iv) The next job is to record the plants which are on the line, but not every plant. If your transect is 100 m long it would take far too long to record all the plants, so instead you take a sample at regular intervals along the line and record the samples only. Look at Fig. 7 to see how this is done. It shows one end of the transect line, with circles and crosses to represent the plants that are on either side of the line. Consider the stake, which is 0 m from the end of the line. The nearest plant to that is a dandelion. In your book write: '0 m − dandelion.' It is a good idea to measure the height of the plant and record that too. Now consider the 1 m mark on the tape. The nearest plant to the mark is a buttercup. In your book write: '1 m − buttercup.' Now proceed to the

---

wood? Does ivy ever grow in a field? Does ivy ever grow in any unshaded place? There are dozens of questions that can be asked about these two quadrats. You may find the answers by reasoning, from a book, or from your other biology work. If not, go *back into the field* to find the answer there. Look at grass plants and see how they grow into a continuous carpet; look for violet plants all over your neighbourhood to see whether or not they ever grow in the open or in a field. Is it that they prefer the shaded, dimmer light of the wood? or is it that they prefer the shelter from wind that the trees provide? Or is it a difference in the soil which makes them woodland plants? Observations *and simple experiments* may help to answer some of these questions.

Lists of animals and descriptions of their lives and habits must also be thought about in this way. In fact, no piece of field work is ever truly finished − it always leads on to something more. So when you have neatly written up the results of your first expedition, remember to LOOK at them, THINK, and GO BACK TO THE FIELD.

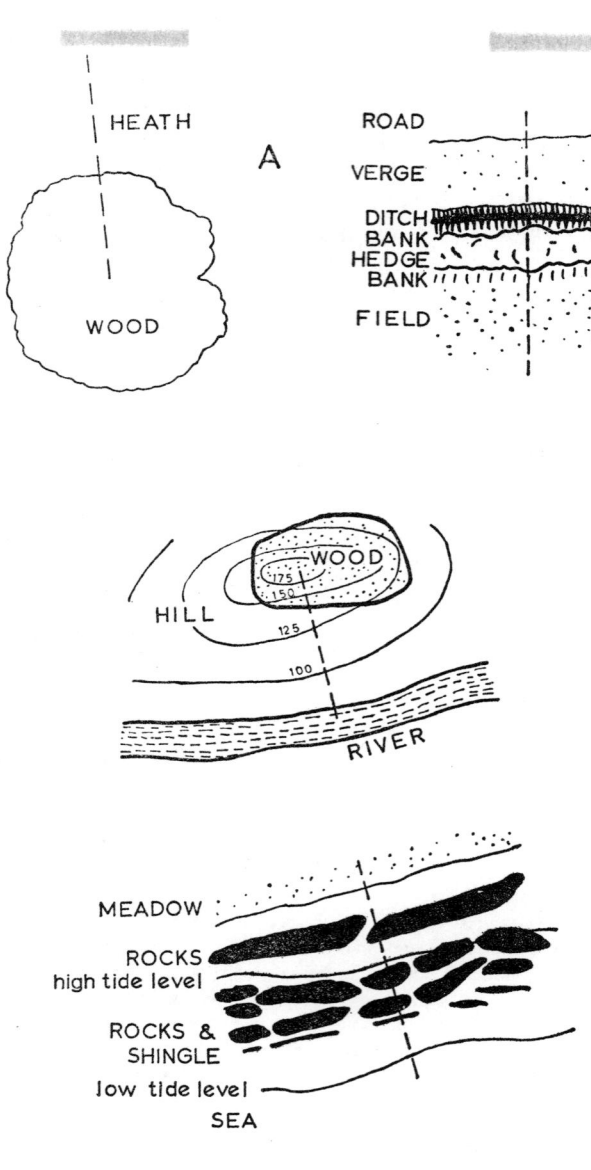

A

HEATH

WOOD

ROAD
VERGE
DITCH
BANK
HEDGE
BANK
FIELD

HILL

WOOD

175

150

125

100

RIVER

MEADOW

ROCKS
high tide level

ROCKS &
SHINGLE

low tide level

SEA

B

FIG. 6. Where to lay transect lines: A, across boundaries between one type of vegetation and another. B, down a slope.

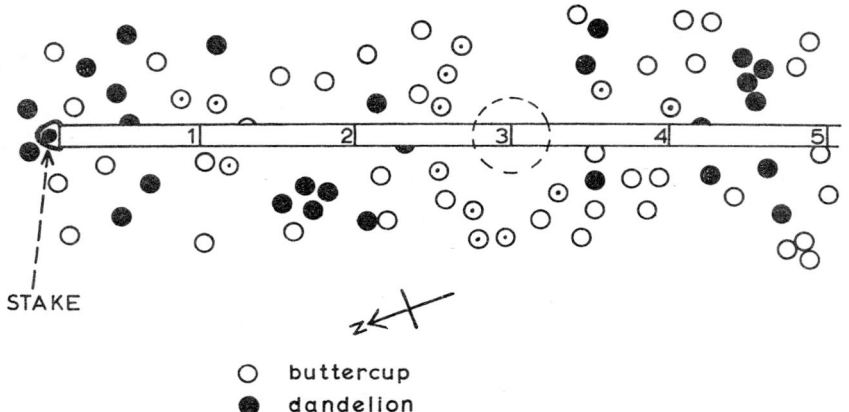

STAKE

○   buttercup
●   dandelion
⊙   shepherd's purse

FIG. 7. How to record plants along a transect line

2 m mark. The nearest plant is a dandelion. Record: '2 m —
dandelion.' At the 3 m mark leave a blank since the nearest
plant is over 10 cm away. Continue this all the way along the
line.

The complete record for the diagram reads:

Distance from end of line (m)		Record	Height (cm)
North-east end	0	D	30
	1	B	65
	2	D	37
	3	—	—
	4	S	21
	5	B	61

KEY

D = dandelion
B = buttercup
S = shepherd's purse
— = bare soil

41

With a fairly short transect, or one that passes through vegetation which changes a lot over a short distance, it is sometimes better to record the plants every 50 cm instead of every metre. If the transect is a longer one you can record every two or three metres. Before you finish make sure that in your records there is some clear indication of which end of the line you started from.

When you are writing your records neatly at home you can, if you wish, simply copy the field records of your transect as they are. This is rather a dull way of presenting the results of so much work and it will be much easier to understand if you combine your transect records into a profile diagram — a sort of 'side view' of the vegetation. An example of this is shown in Fig. 8. The land was level so the soil level is shown by a straight, horizontal line. If your transect is over sloping ground or there are ridges, ditches, and other changes in height, you must take measurements at the same time that you are recording the plants and draw the correct slopes, ridges, and ditches in the profile diagram (see Fig. 9).

In Fig. 8 the plants are represented by symbols. The height of the symbols is in proportion to the height of the plants. At a number of points you will notice that two or even three plants are recorded for one point. At these points the vegetation was very thick and the different plants were so tangled with one another that it was impossible to tell which was the nearest. Also, at these points the plants were to a certain extent 'one on top of another' so in that way both of them, or all three, can be considered as being at one point.

This transect illustrates the effect of tree shade on vegetation. Actually no tree trunk was *on* the line, but their overhanging crowns have affected the plants below them. Along the shady parts of the line (0—27·6 m, 34·8—42·6 m, and 82·8—90·0 m) the main plant is stinging nettle, with dog's mercury and ivy growing beneath it. In the open (28·2—34·2 m, 43·2—68·4 m, and 75·6—82·2 m) the main plants are grasses. There is one part of the line where *both* stinging nettles *and* grasses are growing together (this part runs from about 69 m to about 75 m). This is a region where there was once a tree (the stump of which was found about a metre to one side of the line, level with 72 m). This transect seems to indicate that when the tree was there, stinging

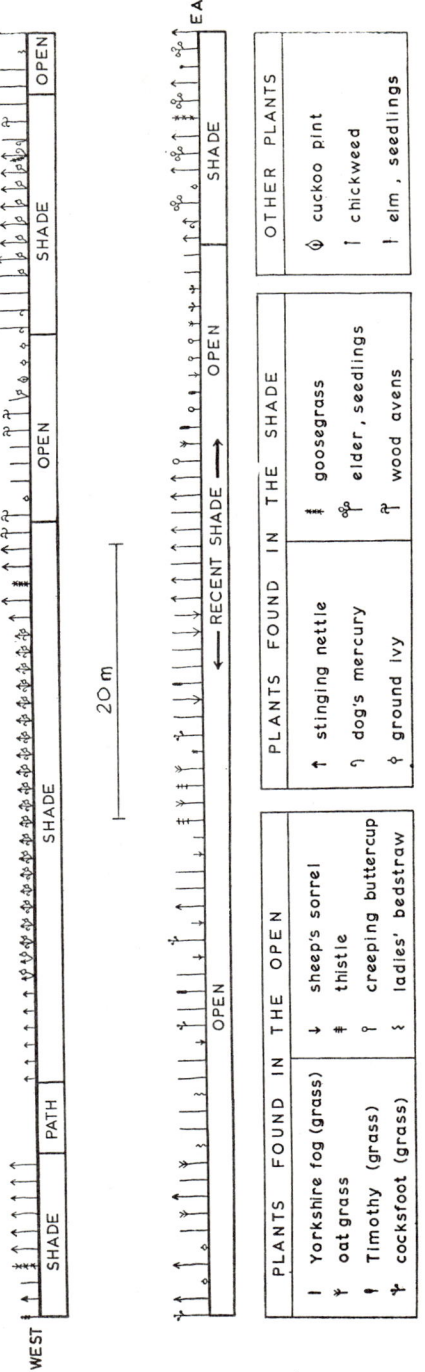

FIG. 8. A transect in open woodland, south of the Mall, Kimbolton Castle, July.
The transect is 90 m long and the plants were recorded every 60 cm.

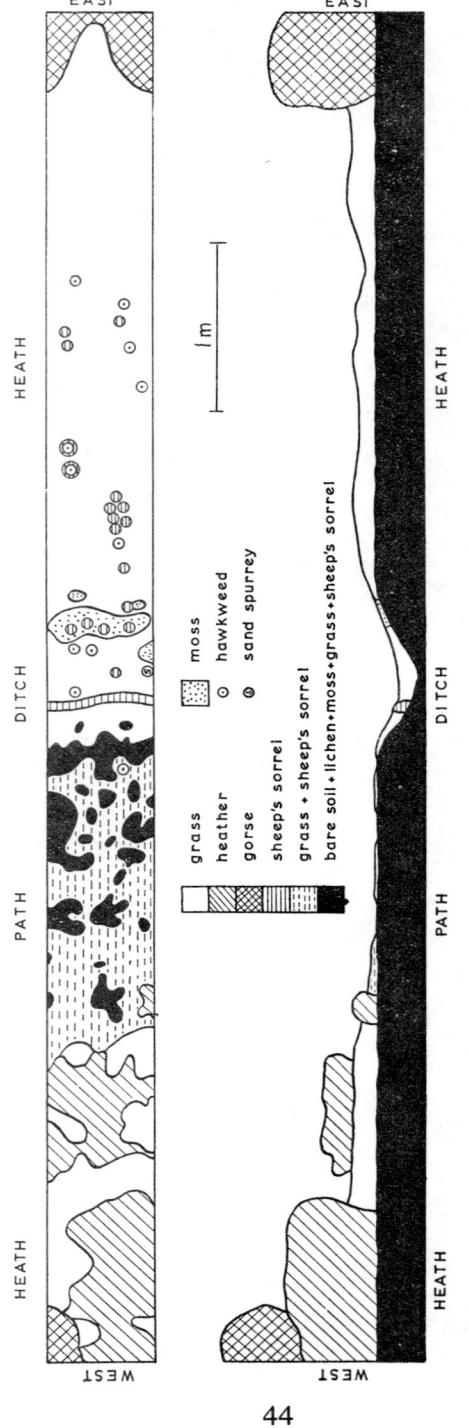

FIG. 9 (*Above*) An example of a belt transect across a footpath on a heath, Worksop College estate, October. The transect consists of thirteen squares, each 60 cm X 60 cm. (*Below*) Profile diagram.

44

nettles grew beneath it but after it fell down or was felled, grass began to grow there instead. The 'shade' vegetation (nettles) is gradually being replaced by 'open' vegetation (grass). It would have been interesting to make further records along this transect a few years later, to see if the grass was increasing and the stinging nettle dwindling. Unfortunately, this area has since been taken under cultivation, and the course of Nature has been interfered with.

Another sort of transect is a belt transect. This is really like a long row of quadrats. An example is shown in Fig. 9. To plot a belt transect, mark the ends and lay the tape as before. Then run a string parallel with the tape and 0·25, 0·5 or 1 m from it (Fig. 10). This will mark out a narrow belt. The vegetation in the belt is plotted in the same way as a quadrat. Using short strings or bamboo canes, mark out the first square and plot it on graph paper. Then move on to the next square and so on.

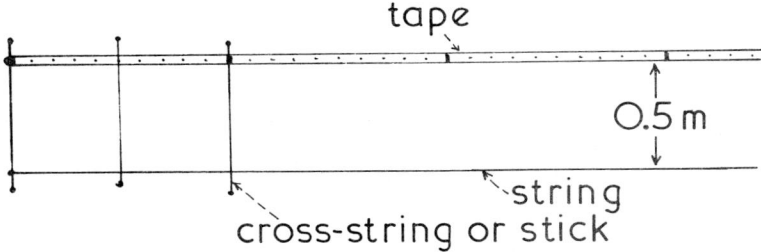

FIG. 10. How to make a belt transect

(c) *Frequency*. It is not necessary to spend a lot of time and effort in making quadrats or transects if you merely wish to say which are the commonest plants in a given area. You can score the commonness or frequency of a plant by using one of the following terms:

a = abundant — one of the commonest plants there
f = frequent — fairly common
o = occasional — not so common
r = rare — very few indeed.

For example, on a heath you might decide that heather and gorse were abundant, grass was frequent, bracken was occasional, and plantains were rare. This is quite a good way of describing the vegetation and it will be useful at times, but it has several disadvantages. Different people will not necessarily agree over the scoring. One may think that heather is abundant, another may say that it is only frequent. Also, showy plants like heather in bloom are so much more obvious to the eye that they appear to be commoner than they really are; yet one may scarcely notice an inconspicuous plant that is really very common. To get a more precise estimate of the relative numbers of different plants, make a *valence analysis*, as follows:

(i) You need a metal ring, 35·6 cm diameter, or a frame 31·6 cm square. These both enclose an area of 0·1 m². One person walks around and suddenly, without looking, throws the ring over one shoulder. His partner watches closely to see where the ring lands (if not, the ring will soon be lost).

(ii) Record the plants that are within the ring where it has landed. This is best done by setting out a table, and ticking the plants which are found inside the ring at each throw, like this:

Number of throw

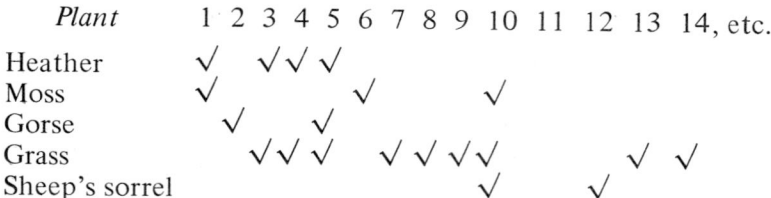

Plant	1	2	3	4	5	6	7	8	9	10	11	12	13	14, etc.
Heather	✓		✓	✓										
Moss	✓					✓				✓				
Gorse		✓		✓										
Grass		✓	✓	✓		✓	✓	✓	✓			✓	✓	
Sheep's sorrel							✓			✓				

In the first throw of this example the ring surrounded heather and moss (the actual number of plants does not matter). The second throw landed the ring on a gorse bush, so gorse was recorded. The third throw landed on grass and heather. This should be continued for at least twenty-five throws or, better still, fifty throws. The example above shows part of the records of a valence analysis of a piece of heathland at Worksop. Altogether thirty throws were made. Out of these, grass was ringed by twenty-four, gorse by nine, sheep's sorrel by seven and heather by only four.

These figures give a better picture of the commonness of these plants on this particular heath. The commonest plant is grass (twenty-four out of thirty). Actually there are many different grasses on this heath but this analysis was made in the winter, when it was not practicable to identify the different sorts, so they are all counted together under the one name 'grass'. Gorse, which one might have said was abundant on this heath, was ringed only nine times, so it is much less common than the grass. Heather was ringed only four times, yet to the eye it appears to be a frequent plant. The analysis shows that it is far less common than sheep's sorrel (seven times) — a plant that one would not notice at all at first glance. This method of estimating the commonness of plants is more reliable than simply looking and estimating and it does not depend on opinions. You should use it whenever you can.

(7) MEASURING LIGHT INTENSITY

Light affects both plants and animals, so it is useful to be able to measure its intensity. Use a photoelectric exposure meter or Pocket Environmental Comparator (Griffin and George, Ltd.).

Some photographic meters are reflected light meters. When used they are pointed at the *subject*. The sort of exposure meter that is built into a camera will be of this type. To use it for measuring the light intensity (or rather, for comparing the light intensity in two places) place a piece of white cardboard, measuring 30 cm square, on the ground. Hold the meter 30 cm above the card, pointing *downward* at its centre and take the reading on the dial of the meter. The white card is a standard object and you are measuring the light being reflected from it under different sorts of illumination.

Other meters are incident light meters and are more convenient to use for this work. The meter has a sheet or cone of white, translucent plastic over its opening and is pointed not at the subject but towards the *camera*, when being used for photography. To use it for comparing light, hold it as near the ground as possible, pointing *upward*, so that light falls on to the white, translucent sheet or cone. Take the reading on the dial.

Readings which are to be compared should be taken within a few minutes of one another if possible, because changes in cloudiness will have big effects. Watch the clouds and if the sun was shining brightly when you took a reading in the open, wait until it is shining again before you take a reading under a tree.

The readings of most meters are figures which are proportional to the intensity of the light. If the reading in one place is twice the reading in another, the light at the first place is twice as strong as the light in the second.

Sometimes shade is patchy because shafts of sunlight shine between the leaves. In this case, take ten readings at different places beneath the tree and work out their average.

(8) MEASURING HUMIDITY

Humidity is, roughly speaking, the 'dampness' of the air. It is expressed as a percentage. Air which is 100 per cent humid contains as much water vapour as it can hold — it is on the point of becoming misty or foggy. Air which is 0 per cent humid is absolutely dry. The humidity of the air affects plants and animals — many small animals cannot live for long if the air is too dry. Measure humidity in one of the following ways:

(a) Use an instrument called a hygrometer. This contains something which changes its size or shape according to the humidity (a human hair is sometimes used). It works like the scales of a fir cone which open out in dry weather and shut when it is wet. The change in size or shape causes a needle to move over a dial and this indicates the humidity of the air. To use this instrument, place it where required, wait until the needle settles, and record the reading.

(b) Another type of hygrometer has two thermometers, one of which has its bulb covered by a cloth bag which is kept wet with water from a small container. This is sometimes called a wet-and-dry-bulb thermometer. Read the temperatures indicated by the two thermometers and then use special tables, which are supplied with the thermometer, to calculate the humidity. The ordinary wet-and-dry-bulb thermometer takes a long time to settle after it has been put in a new place so it is too slow for field work. A whirling hygrometer has the two thermometers fixed to a frame with a handle. The frame and thermometers can be whirled round very

48

quickly — like a rattle. This gives a steady and accurate reading after a minute or two, so it is very suitable for field work. The water reservoir must be kept full.

(9) MEASURING TEMPERATURE
Ordinary laboratory thermometers may be used, but take special care of these in the field, so that they are not broken. Keep them in their protective case when not in use. If you are recording air temperatures you must make sure that direct sunlight does not shine on the thermometer — if it does you will get a reading that is far too high. Hold the thermometer in the shade — if there is no shade about stand with your back to the sun and let your shadow fall on the thermometer.

For special purposes there are several types of thermometer available which can measure maximum temperatures, minimum temperatures, or both. Your teacher will show you how to use these. The Pocket Environmental Comparator has a special temperature probe. This may be attached to the Comparator and used for measuring temperatures in the air (keep it in the shade!), in water (it is waterproof) or in the soil (it is tough and can be buried in soil without breaking).

(10) STUDYING THE SOIL
The nature of the soil affects the plants that grow there and the animals that are found there. Differences in soil will often explain differences between the plants and animals of two areas.

(a) *Soil profile*. This is a section through the soil. Dig a pit about 0·5 m square and as deep as you can manage. One side of the pit should be smooth and vertical. Examine this side and notice the different layers of soil. Measure the thickness of each layer. In your field note-book draw a scale diagram to show these different layers. With a trowel, take a small sample of each layer and examine it. Write a short description of each layer beside the diagram, mentioning its colour and texture (powdery, peaty, sticky, stony, sandy, clay, chalky, rocky — one or more of these words can be used to describe soils), and whether the roots of plants are growing in it. To make your profile complete, take a sample (about two trowelfuls) from each layer, place it in a labelled polythene bag and take it back to the laboratory for testing by the methods described in the next section. The results of these tests can be written on the soil profile diagram, beside each layer.

Another way of displaying the soil profile is to take a strip of plywood about 7 cm wide and 40 cm long. On this strip draw the profile full size. Working on each layer in turn, smear the wood with glue and cover it with some of the soil belonging to that layer. Knock off the surplus soil, leaving the wood covered with a thin coating of soil. When you have finished, your soil profile should look just like the side of the pit you dug. If the soil contains large stones, crack some in half so that they may be stuck to the wood.

(*b*) *Simple soil tests*. Carry out these tests on samples from the layers of a soil profile or on samples dug from the top 25 cm of soil.

TEST 1: MECHANICAL ANALYSIS OF SOIL

Method. Half fill a gas-jar with soil. Add water almost to the top of the jar. Stir the soil vigorously, making sure that none is stuck on the sides or bottom of the jar. Put a lid on the jar and, holding the lid in place, shake the jar several times to mix its contents. Quickly put it down on the bench and leave it to stand for an hour.

Results. The larger particles settle to the bottom first. The finest particles settle last and may take a week or more to do so. After an hour, all but the finest particles will have settled and your jar will look something like Fig. 11. Draw the contents of the jar, making the layers the correct thickness in your drawing. Add notes to your drawing, describing the

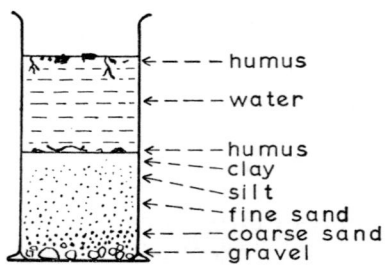

FIG. 11. Mechanical analysis of soil
(Test 1).

appearance of each layer. The humus usually floats on top but after a few days most of it will sink on to the soil.

Repeat this test, using different soils. Which soil contains the most sand? Which has the most clay? Are there any other differences?

TEST 2: TO COMPARE THE DRAINAGE AND WATER-HOLDING PROPERTIES OF TWO SOILS

Method. Leave two samples of soil in dishes in a warm place for a week or dry them for 24 hours in an oven at 100°C. Place each sample in a large funnel, the bottom of which has been loosely plugged with glass wool. The soils should be evenly pressed down and be at the same level in each funnel. Underneath each funnel place a beaker to catch the water when it drains through.

In each hand, take a beaker containing exactly 100 cm³ of water. At exactly the same moment empty one beaker into each funnel while your partner sets a stop-clock going. Make a note of the following results:

Results.

	Soil A	*Soil B*
Time taken for first drop of water to come from funnel	minutes	minutes
Time taken for water to finish draining*	minutes	minutes
Volume of water which drained through	cm³	cm³
Thus volume of water held by soil (100 cm³ − the volume which drained through)	cm³	cm³

* You may consider that the soil has finished draining if 1 minute passes without a drop falling from the end of the funnel.

Interpretations. Which soil drains more quickly? Which soil holds more water?

TEST 3: TO MEASURE THE AIR CONTENT OF SOIL
Method. For this test the soil sample must be taken in a

special way that does not disturb the natural packing. Take an old, used food tin about 8 cm in diameter and 6 cm high. Cut off the rim at the top, leaving a smooth, sharp edge (use tin-snips for doing this), and punch a small hole in the bottom of the tin.

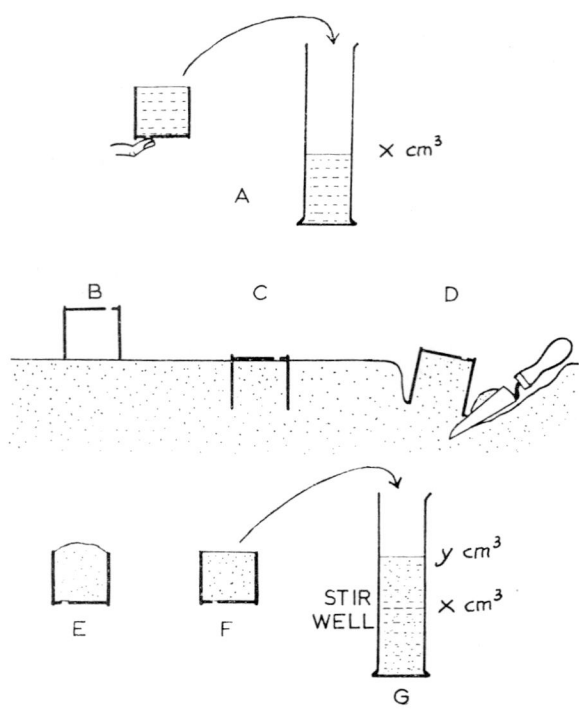

FIG. 12. Stages in Test 3.

Fill the tin to the brim with water, holding your finger over the hole (Fig. 12A), and then pour the water into a 500 cm³ measuring cylinder. Take the reading (x) which gives the volume of the tin.

Place the tin on the soil, the surface of which should be smooth (Fig. 12B). Push the tin into the soil, using a slight twisting motion, until it is full of soil (Fig. 12C). The small hole is to let air escape from the tin should the soil happen to be wet. Dig the tin out of the soil (Fig. 12D), invert it (Fig. 12E), and level off the soil with a straight stick

(Fig. 12F). You now have a quantity of soil equal in volume to the water in the measuring cylinder. The soil has not been compressed or loosened while taking the sample — it is packed just as it was in its natural condition in the soil.

Add the soil to the water in the measuring cylinder — watch *air* bubbling out of it. Stir vigorously to get rid of all air bubbles and pockets of air in the soil. Take the reading which gives the volume (y) of the mixture (Fig. 12G).

Results.

$$\left.\begin{array}{l} \text{Volume of water} = x \text{ cm}^3 \\ \text{Volume of soil (including air)} = x \text{ cm}^3 \end{array}\right\} \quad \begin{array}{l} \text{A tinful of water and the} \\ \text{same tinful of soil were} \\ \text{mixed.} \end{array}$$

Therefore

Volume of water + volume of soil (including air) $= x + x = 2x$ cm^3

But volume of *mixture* of water and soil (air gone) $= y$ cm^3
Therefore

$$\text{Volume of air} = 2x - y \text{ cm}^3$$

Calculate the volume of the air and express it as a percentage of the original volume of the soil:

$$\frac{\text{Volume of air}}{x} \times 100 \text{ per cent}$$

You need to calculate the percentage so that you can compare one soil with another.

TEST 4: TO MEASURE THE AMOUNT OF WATER IN SOIL
Method. Weigh a sample of fresh soil (about 100 g will do). Spread the soil in a shallow dish and put it in an oven at 100°C. On the following day, or later, weigh the soil again. Return it to the oven.

Weigh it again on the following day, or later. If it has not changed its weight since the last time, it is completely dry. If it is still losing weight, return it to the oven. Take it out and weigh it every day until it stops losing weight.

Results.

Weight of fresh soil =   g
Weight of dry soil =   g
Difference = weight of water =   g

Express the weight of water as a percentage of the fresh soil weight:

$$\frac{\text{Weight of water}}{\text{Weight of fresh soil}} \times 100 \text{ per cent}$$

You need to calculate the percentage so that you can compare one soil with another.

With the soil probe attached to the Pocket Environmental Comparator, you can quickly measure and plot the soil moisture levels over an area of soil. The scale readings on the Comparator can be converted to percentages by taking soil samples, measuring their water content, using the Comparator, then performing the above test on each sample to find the percentage of water it contains. You can then plot a graph of scale reading against percentage. This graph can then be used to convert other scale readings into percentages.

TEST 5: TO MEASURE THE AMOUNT OF HUMUS IN SOIL
Method. Take a sample of *dry* soil (about 20 g will do). You can use the dry soil from Test 4. The soil must be dry because it is to be heated to burn off the humus; if there was water in the soil this would go too and the result would show the weight of water and humus *together*.

(*a*) Weigh a crucible.

(*b*) Put the soil in the crucible and weigh again.

(*c*) Heat the crucible and soil strongly over a bunsen burner for 15 minutes. It should become red hot. You will need to handle the hot crucible with crucible tongs.

(*d*) After heating, leave it to cool. Then weigh it.

(*e*) Heat again strongly for a further 15 minutes. Cool. Weigh again.

(*f*) Repeat the heating, cooling and weighing until no more weight is lost.

Results

$$\begin{aligned}
\text{Weight of crucible and dry soil} &= \quad \text{g} \\
\text{Weight of empty crucible} &= \quad \text{g} \\
\text{Difference} = \text{weight of dry soil} &= \quad \text{g}
\end{aligned}$$

$$\begin{aligned}
\text{Weight of crucible and heated soil} &= \quad \text{g} \\
\text{Weight of empty crucible} &= \quad \text{g} \\
\text{Difference} = \text{weight of heated soil} &= \quad \text{g}
\end{aligned}$$

Use the two figures you have just calculated to find the weight of humus:

$$\begin{aligned}
\text{Weight of dry soil} &= \quad \text{g} \\
\text{Weight of heated soil} &= \quad \text{g} \\
\text{Difference} = \text{weight of humus} &= \quad \text{g}
\end{aligned}$$

Express the weight of the humus as a percentage of the dry soil weight:

$$\frac{\text{Weight of humus}}{\text{Weight of dry soil}} \times 100 \text{ per cent}$$

You need to calculate the percentage so that you can compare one soil with another.

TEST 6: TO COMPARE THE PHYSICAL PROPERTIES OF TWO SOILS

Method. Fill two pie-dishes with two different soils. In each dish make a 'mud pie' with water. Mix the soils thoroughly with water, leave them to stand for a few minutes and pour off any water that collects on top. Leave the dishes to stand in a warm place until the soils are dry. It is a good idea to try this experiment using builder's sand in one dish and a heavy clay in the other. This will give you some idea of what to look for when you are testing other soils.

Results. Examine each soil when it is dry. Is the soil loose or is it set into a hard cake? If it is a cake, has the cake shrunk as it dried, or has it cracked into a number of smaller cakes? Run your finger over the surface: is it smooth and yet hard, or is it rough and crumbly?

Interpretations. If you are comparing sand and clay, how do they differ? If you are testing another soil, is it sand-like or clay-like in its properties? How well does this agree with the results of Tests 1 and 2, using the same soil?

TEST 7: TO ESTIMATE THE CALCIUM CARBONATE CONTENT OF SOIL

Calcium carbonate is present in larger amounts in soils from limestone or chalk districts. Some plants grow better in soils with plenty of calcium carbonate and some grow better in soils with little.

Method. Place a little soil in a test-tube and add a few drops of concentrated hydrochloric acid.

Results. If calcium carbonate is present the soil will 'fizz' because of the carbon dioxide produced when the acid acts on the calcium carbonate. If you try this test on two soils at a time you can, by comparing the amount of fizzing, get some idea of which soil contains the most carbonate.

TEST 8: TO MEASURE THE pH OF THE SOIL

pH is a way of describing the degree of acidity or alkalinity of a solution. pH 7 represents a neutral solution (neither acid nor alkaline). A pH less than 7 represents acid conditions (the lower the number, the more acid it is) and a pH greater than 7 represents alkaline conditions (the higher the number the greater the alkalinity). The pH of the soil water or pond and river water can have a big effect on the plants and animals living there.

Method. Special sets of apparatus are sold for measuring pH. The BDH soil testing outfit is a convenient one to use. The Sudbury Soil Test Kit measures not only pH, but also nitrogen, phosphorus and potassium deficiencies, this too is easy to use. Follow the instructions given with the outfit.

Results. Record the pH of the soil you are testing.

(11) IDENTIFYING PLANTS AND ANIMALS

On pp. 59—60 are given the names of some books which will help you to identify all the commoner plants and animals. Simple books are easy to use but they suffer from one dis-advantage — they do not include the less common plants and animals. Quite often you will want to know the name of a less common one. Unless you have access to some of the more complete (and more difficult) books you will have to be content with an approximate identification such as 'a mayfly larva similar to *Cloeon dipterum*' or give its identity only as far as its group: 'a mayfly larva of the *Beetidae*.' If you can, say in what way it differs from the most similar example in your book. If you cannot find anything like your specimen in your book, write an illustrated description of it and leave it at that.

Identification takes a lot of time and, as time can ill be spared on field work, it is useful to have a reference collection of the habitat you are studying. A collection of the twenty or thirty commonest plants on a heath will take much less time to look through than the several hundred plants described in a book. When a habitat is being studied for the first time the plants and some of the animals can be collected and preserved in a reference collection. They will all have to be identified, but only once each. After that, people can use the reference collection instead of having to identify the same plants and animals all over again the following year.

When making quadrats and transects there is the difficult problem of identifying the tiny seedling plants which have no flowers and perhaps only two cotyledon leaves. If you are intending to come back again to the same area, mark the plant in your record as 'seedling X' and then wait for it to grow. When you come back a month or so later it will have grown into a recognizable plant. One book which will help you to identify some of these seedlings is by R. J. Chancellor (see p. 59).

(12) NOTE-BOOK
Keep a field note-book in which you record all your observations whilst you are *in* the field *at the time* the observations are made. Never rely on your memory. If you are intending to record information in the form of a table, draw the table before you set out, with a space for each piece of information you intend to collect. In this way you are less likely to forget to make important observations in the field only to remember them after you have come home again. If you will need graph paper for quadrats, remember to take some with you.

(13) WRITING UP
The results of your field work should be written up neatly in another book as soon as possible. Transfer your results and observations from your field note-book and add any further diagrams, drawings, and observations made in the laboratory. If your work leads to any conclusions, add them to your account.

If a problem is being tackled by a whole class, working in several groups, the results from individual groups can be written on loose leaves and then collected together in a binder to make a class record.

(14) ORGANIZING FIELD WORK

Before you go out into the field decide what you are going to do while you are there. You should write this down in a few sentences. Then make a list of the equipment you will need.

Before you come home again, check through the written 'instructions' you gave yourself at the beginning to see if you have completed your tasks. Then check through the list of equipment to make sure that you take home everything that you brought out.

If a larger project such as a complete survey is being under-taken, the work will be divided among the groups. Someone will have to divide the work and give the instructions to each group, which should then make its own lists of equipment.

BOOK LIST

* Books marked with an asterisk are those which the author has found most useful.

Some books to help you identify and classify plants and animals

G. ALLEN and J. DENSLOW: *Flowerless Plants** (Oxford University Press).

S. ARY and M. GREGORY: *Oxford Book of Wild Flowers** (Oxford University Press).

J. H. BARRETT and C. M. YONGE: *Pocket Guide to the Seashore** (Collins).

F. H. BRIGHTMAN and B. E. NICHOLSON: *Oxford Book of Flowerless Plants** (Oxford University Press).

R. J. CHANCELLOR: *Identification of Weed Seedlings of Farm and Garden** (Blackwell).

J. CLEGG: *Freshwater Life of the British Isles* (Warne).

J. L. CLOUDSLEY-THOMPSON and J. SANKEY: *Land Invertebrates**, a guide to British worms, molluscs, and arthropods (excluding insects) (Methuen).

Clue Book Series (Oxford University Press).

G. B. CORBET: *Identification of British Mammals* (British Museum (NH)).

A. DALE: *Patterns of Life** (Heinemann). Contains many useful identification keys.

A. DARLINGTON: *Pocket Encyclopaedia of Plant Galls in Colour* (Blandford).

R. S. R. FITTER and R. A. RICHARDSON: *Pocket Guide to British Birds** (Collins).

R. S. R. FITTER and R. A. RICHARDSON: *Pocket Guide to Nests and Eggs* (Collins).

M. LANGE and F. B. HORA: *Guide to Mushrooms and Toadstools** (Collins).

T. T. MACAN: *Guide to Freshwater Invertebrate Animals** (Longman).

J. E. MARSON: *Non-aquatic Animal Identification Sheets* (School Natural Science Society publication no. 12).

J. E. MARSON: *Water Animal Identification Sheets* (School Natural Science Society publication no. 8).

W. K. MARTIN: *Concise British Flora In Colour* (Michael Joseph and Ebury Press).

D. McCLINTOCK and R. S. R. FITTER: *Pocket Guide to Wild Flowers** (Collins).

Observer's Books Series (Warne).

K. PAVIOUR-SMITH and J. B. WHITTAKER: *A key to the Major Groups of British Free-living Terrestrial Invertebrates** (Blackwell).

R. PETERSON, G. MOUNTFORT and P.A.D. HOLLOM: *Field Guide to the Birds of Britain and Europe* (Collins).

C. T. PRIME and R. J. DEACOCK: *Trees and Shrubs** (Heffer).

F. ROSE: *Observer's Book of Grasses, Sedges and Rushes* (Warne).

W.J. VAN REINE: *Plants and Animals of Pond and Stream* (Murray).

Wayside and Woodland Series (Warne).

Publications of the School Natural Science Society may be obtained by post from M. J. Wootton, B.Sc., 44 Claremont Gardens, Upminster, Essex, RM14 1DN.

Other books which you will find helpful for field studies

D. P. BENNETT and D. A. HUMPHRIES: *Introduction to Field Biology* (Arnold).

O. N. BISHOP: *Outdoor Biology*, Books 1, 2, and 3 (Murray).

A. DARLINGTON: *Ecology of Refuse Tips* (Heinemann).

A. DARLINGTON: *Natural History Atlas of the British Isles* (Warne).

A. DARLINGTON: *World of a Tree* (Faber).

E. A. R. ENNION and N. TINBERGEN: *Tracks* (Oxford University Press).

M. J. D. HIRONS: *Insect Life of Farm and Garden* (Blandford Press).

M. KNIGHT: *Young Field Naturalist's Guide** (Bell).

M. KNIGHT: *Frogs, Toads and Newts in Britain** (Brockhampton Press).

J. SANKEY: *A Guide to Field Biology** (Longman).

T.R.E. SOUTHWOOD: *Life of the Wayside and Woodland* (Warne).

J. O. L. SPOCZYNSKA: *Practical Field Work for the Young Naturalist* (Muller).

J. O. L. SPOCZYNSKA: *Zoo on your Window Ledge** (Muller).

A. L. WELLS: *Microscope Made Easy** (Warne).

KEY TO COMMON YELLOW DANDELION-LIKE FLOWERS

This key is to help you with on-the-spot identifications of the many dandelion-like flowers you will come across in your field work.

1   No leaves on flower stems (but there may be small leafy scales).   2
Flower stems have normal leaves.   9

2   Flower stems have many purplish scale-leaves; stems have woolly hairs; the real leaves are large and rounded and appear long after the flowers.
        COLTSFOOT *(Tussilago farafara)*
Not as above.   3

3   Leaves with long stiff white hairs.
        MOUSE-EAR HAWKWEED *(Hieracum pilosella)*
Leaves not as above.   4

4   Flower stems usually branched; or unbranched with small leafy scales scattered along the stem.   5
Flower stems unbranched; if there are leafy scales, they are found only just below the flowers.   8

5   'Flower' has many narrow scales among the yellow florets.
        CATSEAR *(Hypochaeris radicata)*
No scales.   6

6   'Flower' surrounded by two rows of greenish scales, the outer row shorter than the inner row.
        SMOOTH HAWKSBEARD *(Crepis capillaris)*
'Flower' surrounded by more than two rows of scales.   7

7 Flower stem has many green leafy scales just below each flower.
> AUTUMNAL HAWKBIT *(Leontodon autumnalis)*

Not as above.
> HAWKWEEDS *(Hieracium* species)

8 Flower stems hollow, smooth and shiny; milky juice appears when stems are broken
> DANDELIONS *(Taraxacum* species)

Not as above
> ROUGH HAWKBIT *(Leontodon hispidus)*

9 Leaves at base of stem, long and narrow, like leaves of grass; 'flowers' close at midday.
> GOAT'S-BEARD/JACK-GO-TO-BED-AT-NOON *(Tragopogon pratensis)*

Not as above.   10

10 'Flower' surrounded by two rows of greenish scales, the outer row shorter than the inner row.   11

'Flower' surrounded by more than two rows of scales.   15

11 Fruits have no tufts of hairs on them.
> NIPPLEWORT *(Lapsana communis)*

Fruits have tufts of white hair or brown hair on them.   12

12 Greenish scales around the 'flower' make a long narrow tube; usually only five yellow florets.
> WALL LETTUCE *(Mycelis muralis)*

Greenish scales make broad tube or 'cup'; more than five florets.   13

13 Hairs on fruit are at the end of a long thin stalk (like a dandelion fruit).
> BEAKED HAWKSBEARD *(Crepis taraxacifolia)*

Not as above.   14

14 Hairs on fruit are brittle and brownish. Found
in wet meadows from Worcestershire north-
wards.

> MARSH HAWKSBEARD *(Crepis paludosa)*

Not as above.

> SMOOTH HAWKSBEARD *(Crepis
> capillaris)*

15 Leaves and leaf-stalks hairy, but never prickly;
when stems are broken milky juice is not easily
seen.

> HAWKWEEDS *(Hieracium species)*

Leaves and leaf-stalks with no hairs or very
few; leaves are often prickly on the edges and
on the underside of the main vein; when stems
are broken much milky juice is seen.   16

16 Tall plant (1-2m) with creeping underground
stems. 'Flower' large (3-5cm diam.).

> FIELD SOW-THISTLE OR MILK-
> THISTLE *(Sonchus arvensis)*

Not as above.   17

17 The 'ears' at the base of each stem leaf are
rounded and pressed close to the stem; leaves
are usually very prickly.

> PRICKLY OR SPINY SOW-THISTLE OR
> MILK-THISTLE *(Sonchus asper)*

The 'ears' are pointed and not close to the
stem; leaves are not prickly.

> SMOOTH SOW-THISTLE OR MILK-
> THISTLE *(Sonchus oleraceus)*

Now check your identification against a picture or
detailed description in a book about wild plants.